Daniel Webster Mead

Notes on the Selection and Design of Public Works

Hydraulic and Power Plants, Electric Lighting and Pumping Installations,

Water Works, Sewerage and Drainage

Daniel Webster Mead

Notes on the Selection and Design of Public Works
Hydraulic and Power Plants, Electric Lighting and Pumping Installations, Water Works, Sewerage and Drainage

ISBN/EAN: 9783337269715

Printed in Europe, USA, Canada, Australia, Japan

Cover: Foto ©berggeist007 / pixelio.de

More available books at **www.hansebooks.com**

NOTES

ON THE

SELECTION AND DESIGN

OF

PUBLIC WORKS

HYDRAULIC AND POWER PLANTS
ELECTRIC LIGHTING AND PUMPING INSTALLATIONS
WATER WORKS, SEWERAGE AND DRAINAGE

BY

DANIEL W. MEAD, CONSULTING ENGINEER

MEMBER

AMERICAN SOCIETY CIVIL ENGINEERS
AMERICAN INSTITUTE MINING ENGINEERS
AMERICAN WATER WORKS ASSOCIATION
AMERICAN PUBLIC HEALTH ASSOCIATION
ILLINOIS SOCIETY ENGINEERS AND SURVEYORS
WESTERN SOCIETY ENGINEERS
FRANKLIN INSTITUTE

FIRST NATIONAL BANK BUILDING

CHICAGO, ILL.

Contents.

	PAGE.
Introduction,	7
Development and Use of Power,	8
Equivalent Units of Power, Energy and Work,	10
Duty,	11
Duty Table,	12
Duty and Efficiency of Pumping Machinery,	13
Examination and Testing of Power Plants,	14
The Development of Energy from Fuel,	15
The Steam Engine,	16
Indicator Diagrams,	17
Heat Engines,	18
Electrical Energy,	20
Cost of Street Lighting,	21
Examination and Valuation of Public Works,	22
Energy Lost in Electrical Transmission,	23
The Switchboard,	24
The Lawrenceville Water and Light Plant,	26
The De Kalb Electrical Pumping Plant,	30
The Examination and Improvement of Water Supplies,	32
The New Water Supply System at Rockford, Illinois,	33
Water Storage,	40
Drainage,	45
The Hydraulic Power Plant of the Rockford General Electrical Company,	48
Hydraulic Machinery,	52
Foundation Work,	54
Sewers and Sewerage,	56
Special Machinery for Construction,	57
Notes on Pumps and Pumping,	58
Hydraulic Tables,	59
Fire Streams,	62
Finale,	63

Introduction

The officials of municipal or private corporations who find it necessary to undertake important improvements, such as the installation of water works, sewerage systems, electric lighting plants or power plants for various purposes, are brought at once face to face with complex and important problems. Invention has made manifold methods available for each case, and interested parties stand ready to prove (to their own satisfaction at least) the pre-eminent value of the goods or methods they represent. The conflicting evidence is confusing to those unacquainted with the details.

While there may be many methods and plans for construction of an improvement, there can be but one *best plan*, and that plan can only be determined by judgment based on extended experience. It is safe to say that a very large proportion of the public works of today are needlessly expensive. Sometimes they may be cheap in first cost; too cheap, indeed, for ultimate economy; and while much attention is commonly given to this point, the cost of maintenance and operation is often practically neglected. The cost of operating a large percentage of the water, light and power plants now in use, could be materially reduced by the adoption of proper machinery and methods. Economy should include all phases of the question.

It is important, when new works are to be installed, to know what has been done in similar cases elsewhere. It is hoped that the following notes, data and illustrations from the writer's practice, will be of interest and value in this connection. The writer will be glad to furnish further information concerning the works in question or to design and superintend the construction of similar works. New plants will be designed and installed, and old plants remodeled on an economical and modern basis. Attention will also be given to the testing of water, light and power plants, and to the examination and reports on their condition and value.

DANIEL W. MEAD,
First National Bank Building,
CHICAGO.

Development and Use of Power.

The development of power from fuel, from water or from other natural sources of potential energy and the transformation of such energy into forms which can be utilized for commercial purposes, constitutes a large factor in the development of public and private engineering installations. The annual expenditure of energy and consequently of capital in one form or another, is so large that the question of economy in power development, transmission and utilization demands the most careful consideration and investigation.

Energy is the active principle of the universe. It is the basis of all action and of all physical phenomena. It is the ability to exert force, to do work. It is manifest in various forms. The energy of steam transports commerce and drives factories; as water power it turns the mill wheel, as electricity it propels cars, lights streets and houses, and as wind it drives the wind mill and propels the ship.

Work is the application of energy to particular purposes. It is the exertion of force through space. The unit of work is the foot pound, or the equivalent of the amount of work required to raise one pound one foot; one pound raised one foot: .1 pound raised ten feet: ten pounds raised .1 foot, or any subdivision of pounds and feet whose product will equal one requires one foot pound of work to perform it.

Power is the rate at which work is accomplished, or the amount of work done in a given time.

The unit of power is based on the unit of work and is called the horse power. It is work performed at the rate of 550 foot pounds per second or 33,000 foot pounds per minute.

The unit of heat is the amount of heat which will raise one pound of distilled water from 39° to 40° Fahr. It is called the British Thermal Unit and is indicated by the initials B. T. U.

The unit of quantity of electricity is the coulomb. One coulomb per second is designated an ampere and one ampere under a volt pressure is a watt. The watt and kilowatt (or 1,000 watts) are the units of electrical power.

Water power is the power obtained from falling water and the units may be the gallon or cubic foot for quantity and the foot head or pounds pressure per square inch for weight with the second or minute for time.

Definite quantities of work are also designated by the "horse power hour" equivalent to 1,980,000 foot pounds, and the "kilowatt hour" equivalent to 2,654,150 foot pounds. The unit of work in steam power in ordinary use is the pound of steam with its designated pressure and rate of use. It is based essentially on the heat unit. The pound of steam may be considered as containing an average of 1,000 B. T. U. which may be utilized for power. This is equivalent to 772,000 foot pounds.

Physicists have found that heat, light, electricity, chemical action and all other physical manifestations, are forms of energy. These various forms are in many cases convertible one into the other in certain definite ratios. These ratios, shown in the following tables, are only obtainable by the most careful laboratory methods and cannot be attained in practice on account of heat radiation, leakage, friction, etc. While these losses cannot be wholly avoided, in good practice they must be reduced to a minimum. The ratios given are therefore the ultimate end which the best engineering endeavors to produce in actual practice.

EQUIVALENT UNITS OF POWER AND ENERGY.

Work		Heat	Electricity	Water Power			
Foot Pounds per Minute	Horse Power	Thermal Units per Minute B. T. U.	Watts	Gallon Feet per Minute	Cubic feet head per Minute	Gallons lbs. pressure per Minute	Cubic feet lbs. pressure per Minute
1.	.0000303	.001295	.0226	.12	.016	.0519	.0069
33000	1.	42.746	746	3960	528.	1713.4	229.05
772.	.0234	1.	17.45	92.64	12.352	40.083	5.358
44.24	.00134	.0573	1.	5.308	.70895	2.296	.307
8.34	.00025	.0108	.18356	1.	.1337	.433	.0579
62.396	.00189	.0808	1.4105	7.48	1.	3.24	.433
19.26	.00058	.0249	.435	2.31	.309	1.	.1337
144.08	.00436	.1866	.0326	.00173	2.31	7.48	1.

EQUIVALENT UNITS OF WORK.

	Heat	Electricity	Steam Power Pounds per hour working between limits given below					
Horse Power Hours	B. T. U. per Hour	Kilowatt Hours	From Water at a Temperature Fahr. of					
			212°	212°	60°	296°	60°	200°
			To Steam at a Gauge Pressure of					
			0 lbs.	81 lbs.	100 lbs.	100 lbs.	150 lbs.	150 lbs.
1.	2564.76	.746	2.65	2.545	2.29	2.5	2.18	2.47
.0004	1	.00029	.001035	.001	.00086	.00098	.000859	.000975
1.34	3410.28	1.	3.56	3.41	2.95	3.35	2.92	3.32
.353	966	.281	1.	.996	.87	.95	.855	.94
.394	1080	.293	1.035	1.	.86	.984	.858	.974
.455	1157	.339	1.19	1.16	1.	1.135	.99	1.122
.401	1017	.298	1.05	1.02	.88	1.	.87	.99
.459	1166	.342	1.21	1.17	1.01	1.145	1.	1.12
.403	1025	.301	1.06	1.025	.89	1.01	.88	1.

Duty.

Beside the units of Power, Energy and Work included in the preceding tables, from the relations of which the efficiency of various machines for transforming energy can be determined, there is also another measure of efficiency which is largely used in considering pumping plants and pumping machinery. This is termed Duty and represents the ratio of work done to energy expended in doing it.

Duty is the ratio of the millions of foot pounds of work done to the coal, or steam, or heat units used as power and is usually expressed as million foot pounds duty per 100 pounds coal, per 1,000 pounds dry steam or per 10,000 heat units.

Duty based on coal is very indefinite, for coal varies largely in the potential energy or calorific power which it contains per pound. When coal is used as a basis the plant efficiency is involved including boilers, steam piping, boiler feed pumps, etc., the efficiencies of which do not necessarily have any relation to the individual efficiency of the pump itself.

Duty based on coal should therefore only be used where the entire plant is considered and when the class of coal is also specified.

Duty based on steam used is more specific but hardly sufficiently so, as for example the energy value of steam at 150 pounds gauge pressure is from 16% to 18% greater than that of steam at 90 pounds pressure. Entrained water and condensation in the piping also modify the results so that in using steam as a basis of duty, dry steam at a given pressure should be specified.

Duty based on heat units delivered to the engine while still more specific should also have the steam pressure specified.

In the table following the relation of duty to coal consumption and steam consumption per horse power per hour is shown.

11

Table Showing Duty, Corresponding Amount of Coal per H. P. per Hour and Corresponding Amount of Coal Required to Raise One Million Gallons of Water 100 ft. High.

Duty.	Coal* per H. P. per hour.	Lbs. per million gal. 100 ft. high.	Duty.	Coal* per H. P. per hour.	Lbs. per million gal. 100 ft. high.	Duty.	Coal* per H. P. per hour.	Lbs. per million gal. 100 ft. high.
1	198.00	83398	51	3.88	1635	101	1.96	825
2	99.00	41699	52	3.80	1604	102	1.94	817
3	66.00	27799	53	3.73	1573	103	1.92	809
4	49.50	20849	54	3.66	1544	104	1.90	802
5	39.60	16679	55	3.60	1516	105	1.89	794
6	33.00	13899	56	3.53	1489	106	1.87	786
7	28.29	11914	57	3.47	1463	107	1.85	779
8	24.75	10424	58	3.41	1437	108	1.83	772
9	22.00	9266	59	3.35	1414	109	1.82	765
10	19.80	8340	60	3.30	1389	110	1.80	758
11	18.00	7581	61	3.24	1367	111	1.78	751
12	16.50	6950	62	3.19	1345	112	1.77	744
13	15.23	6415	63	3.14	1323	113	1.75	738
14	14.14	5957	64	3.09	1303	114	1.74	731
15	13.20	5560	65	3.04	1283	115	1.72	725
16	12.37	5212	66	3.00	1263	116	1.71	719
17	11.64	4906	67	2.95	1244	117	1.69	713
18	11.00	4633	68	2.91	1226	118	1.68	707
19	10.42	4389	69	2.87	1208	119	1.66	701
20	9.90	4170	70	2.83	1191	120	1 65	695
21	9.43	3971	71	2.79	1174	121	1.64	689
22	9.00	3791	72	2.75	1158	122	1.62	683
23	8.60	3626	73	2.71	1142	123	1.61	678
24	8.25	3475	74	2.67	1127	124	1.60	672
25	7.92	3336	75	2.64	1112	125	1.58	667
26	7.61	3208	76	2.60	1097	126	1.57	662
27	7.33	3089	77	2.57	1083	127	1.56	656
28	7.07	2978	78	2.54	1069	128	1.55	651
29	6.83	2876	79	2.50	1055	129	1.53	646
30	6.60	2780	80	2.47	1042	130	1.52	641
31	6.38	2690	81	2.44	1029	131	1.51	636
32	6.18	2606	82	2.41	1017	132	1.50	632
33	6.00	2527	83	2.38	1004	133	1.49	627
34	5.82	2453	84	2.36	992	134	1.48	622
35	5.65	2383	85	2.33	981	135	1.47	618
36	5.50	2316	86	2.30	969	136	1.46	613
37	5.35	2254	87	2.28	958	137	1.45	609
38	5.21	2194	88	2.25	947	138	1.43	604
39	5.07	2138	89	2.22	937	139	1.42	600
40	4.95	2085	90	2.20	926	140	1.41	595
41	4.83	2034	91	2.18	916	141	1.40	591
42	4.71	1985	92	2.15	906	142	1 39	587
43	4.60	1939	93	2.13	896	143	1.38	583
44	4.50	1895	94	2.11	887	144	1.37	579
45	4.40	1853	95	2.08	878	145	1 37	575
46	4.30	1813	96	2.06	868	146	1.36	571
47	4.21	1774	97	2.04	859	147	1.35	567
48	4.12	1737	98	2.02	851	148	1.34	563
49	4.04	1702	99	2.00	842	149	1.33	560
50	3.96	1668	100	1.98	834	150	1.32	556

*For corresponding weight of steam per horse power per hour, multiply the figures in this column by 10; and for corresponding heat units, multiply by 10,000.

Duty and Efficiency of Pumping Machinery.

From the data on page 9 it will be seen that if all of the energy of steam were utilized a perfect steam pumping engine should give a duty of 772 million foot pounds for 1,000 pounds of dry steam.

In practice it is found that the various types of direct acting steam pumping engines will give duties about as follows:

Type of Pump.	Duty in million foot lbs. per 1,000 lbs. dry steam.	Corresponding Steam per A. H. P. per hour, lbs.
High duty engines	100 to 160	19.8 to 12.3
Pumping engines	75 " 100	26.4 " 19.8
Large size well designed steam pumps	20 " 40	99.0 " 49.5
Ordinary well designed steam pumps	10 " 20	198.0 " 99.0
Direct acting deep well pumps	2 " 6	990.0 " 330.0

The efficiency of any pump or other machine is the ratio between the power furnished to it and the power actually utilized in work done.

Well designed centrifugal pumps will give under high lifts from 60 to 70 per cent. of efficiency.

Well designed power pumps will give from 65 to 70 per cent. efficiency.

The duty that may be developed with such pumps will depend on the efficiencies of the pumps themselves, and the steam consumption of the engines used to operate them. (See page 16.)

Under ordinary service the air lift pump will give from 15 to 25 per cent. efficiency.

With various types of compressors the corresponding duty would therefore be about as follows:—

Type of Compressor.	Steam consumption lbs. dry steam per I. H. P.	Duty in million foot pounds with 15 per cent. efficiency.	25 per cent. efficiency.
Compound Corliss compressor	16 to 20	19 to 30	31 to 25
Simple condensing Corliss compressor	22 " 28	13 " 10	22½ " 18
Simple Corliss compressor	35 " 40	9½ " 8½	14 " 12
Well designed high pressure compressor	40 " 60	8½ " 5	12 " 8
Small straight line compressor	50 " 80	6 " 4½	10 " 6

Examination and Testing of Power Plants.

The writer will give especial attention to the examination. test and improvement of power plants of all classes. He is provided with the latest and most modern appliances for measuring and indicating the power and efficiency of hydraulic plants, steam engine and boiler plants, heat engines and electric generators and motors. He will undertake to examine power plants and advise as to the possibility and cost of improvements in their efficiency and economy of operation, and will, if desired, give personal and detailed supervision to such construction.

Correspondence concerning installations of which investigations are desired, is requested.

The Development of Energy from Fuel.

Fuel is the source of potential energy most widely used commercially. From wood, coal, petroleum, natural gas and other fuels energy is developed in the form of heat by combustion.

Fuel energy is most commonly utilized by means of the steam boiler.

On account of heat lost in the waste gases from the boiler furnace, only about 83 per cent. of the calorific value of the fuel can be made available. The best boilers will utilize about 90 per cent. of this available energy or about 75 per cent. of the full calorific power of the fuel. With poor boilers often not more than 50 per cent. of the calorific power of the fuel is utilized.

The greatest care is necessary in the design and construction of furnace, boiler and accessories in order to develop maximum efficiencies and secure the most economical results in the utilization of fuels.

CALORIFIC VALUES OF VARIOUS FUELS.

Fuel.	Average Heat Units.		Equivalent evaporation from and at 212° Fahr. lbs	Equivalent horse power hours.
	Per lb.	Per 1,000 cub. feet.		
Coke	14,880	....	15.40	5.84
Anthracite coal	14,660	....	15.17	5.76
Bituminous coal	12,740	....	13.18	5.00
Wood	7,740	....	8.01	3.04
Petroleum	19,150	....	19.82	7.12
Natural gas		885,880	917.06	348.08
Coal gas		570,900	599.93	224.32
Water gas		253,100	262.00	99.43
Producer gas		111,190	115.10	43.69

The Steam Engine.

Of the energy delivered to the engine the proportion actually utilized depends on the character of the engine used, its design and the condition in which it is maintained.

The approximate average steam consumption per indicated horse power per hour of various classes of steam engines is as follows:

Triple expansion condensing Corliss, -	12 to 14 lbs.
Compound condensing Corliss, -	14 to 18 lbs.
Simple condensing Corliss, - -	18 to 21 lbs.
Compound Corliss, - - -	18 to 21 lbs.
Compound condensing Automatic, -	17 to 24 lbs.
Simple Corliss, - - -	24 to 30 lbs.
Simple High Speed, - - -	30 to 36 lbs.
Simple Slide Valve, - - -	33 to 45 lbs.

From 6 to 15 per cent. of the I. H. P. of the engine is lost in friction in well designed engines at full load. At partial load the percentage of loss is much greater.

A perfect engine could utilize only 25 per cent. of the energy of the steam delivered to it. In actual practice, however, the best engines only utilize about 17 per cent. while the ordinary slide valve engine will utilize only about 5 per cent. Poor engines in poor condition will utilize still less, frequently amounting to less than 1 per cent.

Care should be taken to keep an engine in good order and its valves properly set. Improperly set valves are a constant source of loss, to avoid which, the indicator should be frequently applied. The indicator diagrams on the following page show the characteristic effects of common defects.

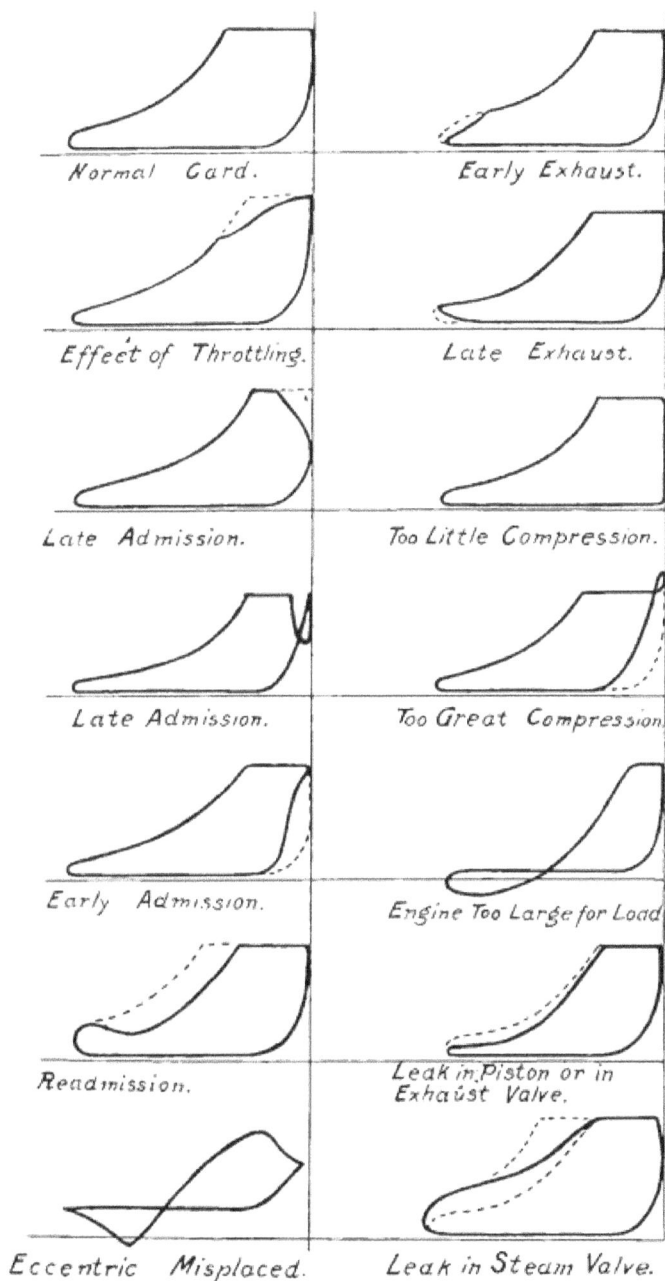

Normal Card.

Early Exhaust.

Effect of Throttling.

Late Exhaust.

Late Admission.

Too Little Compression.

Late Admission.

Too Great Compression.

Early Admission.

Engine Too Large for Load.

Readmission.

Leak in Piston or in Exhaust Valve.

Eccentric Misplaced.

Leak in Steam Valve.

Heat Engines.

Only about 12 per cent. of fuel energy is utilized in the indicated horse power of the best steam engines, while in ordinary practice only from 1 to 3 per cent. is so utilized. As this loss is largely due to the nature of steam, it has resulted in attracting the attention of inventors to other forms of heat engines for power purposes. The best-known forms of these are the various gas and gasoline engines, in which a mixture of air and gas or vapor is ignited or exploded in the engine cylinder itself, without the interposition of a boiler. These engines utilize from 16 per cent. to 20 per cent. of the calorific value of the fuel used. They are also immediately available for power without the slow process of getting up steam. Another favorable condition is the small amount of attention necessary in their operation.

Their availability is entirely a matter of condition, which may be adverse or favorable in any locality or for any purpose. With proper care and under proper conditions, they are very satisfactory. The writer has installed a large number of these engines for various purposes. The illustration on the opposite page shows one of two direct connected engines and pumps installed in the water works at Dundee, Ill. Most of the water for this place is pumped by hydraulic power, the engine and pump being reserved principally for fire service. Full pressure can be furnished by either or both units in 'less than two minutes.

Gasoline Pumping Plant—Dundee, Illinois.

Electrical Energy.

The utilization of energy in electrical transmission, electrical traction, electric lighting and electric power installations for pumping and other commercial purposes is widely extending the application of this source of power. In considering the comparative economy of installations of this kind, the kilowatts developed per pound of coal are often used for comparison. A pound of average coal may be considered as equivalent to 12,000 heat units, which in turn is equivalent to 3,488 watt hours. In actual practice, however, the best electric lighting plants have only been able to generate about 265 watt hours per pound of fuel, while a large percentage of such plants develop less than fifty watt hours per pound of fuel. This indicates a very large opportunity for improvement in the average electric installation. In electric lighting, the amount of current consumed by various methods of installation, is about as follows:

POWER REQUIRED TO OPERATE ELECTRIC LIGHTS.

Candle Power		Kind of Lamp	Kind of Current	Power Required			Watts per actual candle power
Rated	Actual			Amperes	Volts	Watts	
2000	346	Open Arc	Constant Current	10	45	450	1.3
2000	346	Open Arc	Constant Potential	10	55	550	1.59
1200	208	Enclosed Arc	Constant Potential	6	110	660	3.18
1200	208	Enclosed Arc	Constant Potential	3	220	660	3.18
16	16	Incandescent	Constant Potential	.5	110	55	3.4
16	16	Incandescent	Constant Potential	.3	220	66	4.1
16	16	Incandescent	Alternating	2	50	*100	6.25

*Including ordinary transformer losses.
†Two in series on 110 volt circuit.

In practice it may be remembered that one H. P. of the engine will operate from 1 to 1¼ 2,000 c. p. arc lamps and from 8 to 10 16 c. p. incandescent lights.

AVERAGE COST OF ARC STREET LIGHTING IN EACH STATE.

From *Engineering News*.

State	Hours per year	Coal per ton	Cost per lamp per year	Cost per lamp per hour	State	Hours per year	Coal per ton	Cost per lamp per year	Cost per lamp per hour
Alabama	4,000	$1.92	$89.00	$0.028	Nevada	3,000	$9.50	$144.00	$0.050
Arizona	3,596	2.50	121.00	.034	New Hampshire	2,583	4.42	81.50	.030
California	2,585	7.00	119.08	.045	New Jersey	3,618	2.70	105.40	.032
Colorado	3,415	3.18	129.31	.042	New Mexico	3,275	3.92	123.33	.045
Connecticut	2,730	4.28	86.92	.035	New York	3,307	2.87	93.61	.029
Delaware	4,083	2.32	110.00	.028	North Carolina	3,572	3.50	97.50	.021
Dist. of Columbia	4,000		106.00	.027	North Dakota	4,000	6.38	120.00	.030
Florida	2,174		117.05	.054	Ohio	3,350	1.96	78.87	.025
Georgia	3,434	1.36	83.51	.034	Oregon	3,650		127.71	.035
Idaho	4,000		114.00	.028	Pennsylvania	3,201	1.96	88.75	.022
Illinois	2,589	1.19	77.92	.031	Rhode Island	2,912	3.85	102.27	.037
Indiana	2,682	1.79	79.33	.033	South Carolina	3,088	2.68	83.50	.031
Iowa	2,879	1.57	88.73	.032	South Dakota	3,690	5.50	125.00	.034
Kansas	2,254	2.02	97.19	.049	Tennessee	4,000	2.08	88.05	.022
Kentucky	3,061	1.23	88.07	.032	Texas	3,472	3.13	108.75	.037
Louisiana	4,000	2.58	127.50	.032	Utah	3,545	2.63	74.90	.023
Maine	2,621	3.03	67.50	.031	Vermont	2,284	4.37	84.97	.039
Massachusetts	2,705	4.05	93.51	.035	Virginia	3,725	3.07	76.09	.022
Maryland	4,000	3.01	127.33	.032	Washington	3,479	2.75	127.54	.039
Michigan	3,465	2.70	75.04	.024	West Virginia	3,089	.90	79.21	.027
Minnesota	3,333	3.41	95.00	.036	Wisconsin	3,057	3.13	83.16	.030
Mississippi	2,179	2.33	102.00	.047	Wyoming	3,545	1.58	166.00	.052
Missouri	2,917	2.54	83.00	.032					
Montana	4,000	4.00	130.80	.033	Total average, forty-seven states	3,326	3.03	101.18	.034
Nebraska	2,894	2.23	109.50	.041					

COST OF GAS STREET LIGHTING.

Prepared by the League of American Municipalities, 1897.

Place	Price per lamp per year	Life of contract years	Candle power	Schedule	Posts and lamps owned by	No. of gas lights for sts.	Price of gas per M to private consumers
Atlantic City, N. J.	$22.00	Ex.	22	A. N.	City	137	$1.50
Baltimore, M. D	* 1.25	N. C.		A. N.	City	5,151	1.25
Buffalo, N. Y	14.67	5	18	A. N.	City	5,568	1.00
Bridgeport, Conn	18.00	1	16	A. N.	City	75	1.25
Carthage, Mo.	30.00	20	14	M.	Co.	90	1.50
Charleston, S. C.	16.62	1	16	M.	Co.	56	1.75
Dayton, Ohio	19.00	Ex.	18	A. N.	City	1,250	1.00
Erie, Pa	16.00	5	18	A. N.	City	600	1.25
Elmira, N. Y	* 1.70	Ind.		M.	City	49	2.00
Fort Smith, Ark	27.50	15	14	M.	Co.	205	2.50
Grand Rapids, Mich.	29.00	1	60	A. N.	City	50	1.00
La Fayette, Ind.	20.00			A. N.	City	18	1.25
Louisville, Ky	17.47		16	A. N.	Co.	69	1.35
Mt. Vernon, N. Y.	21.50	5	18	A. N.	Co.	565	1.50
Maysville, Ky	25.00	5		M.	C. & Co.	125	1.50
New Haven, Conn	22.25	3	18	A. N.	City	743	1.25
Newton, Mass	16.50	3	18	Mid.	City	948	1.35
Niagara Falls, N. Y.	30.00	N. C.		A. N.	City	50	2.00
New Bedford, Mass	† .65½	1	16	M.	City	480	1.50
Nashville, Tenn.	.75	Ind.	18	M.	City	405	1.40
Providence, R. I.	28.00	3	18	A. N.	City	700	1.10
St. Paul, Minn.	23.00	1	18	A. N.	Co.	2,062	1.30
Sioux City, Iowa	30.00	5	20	A. N.	Co.	52	1.30
Vincennes, Ind	26.00	25	14	M.	Co.	232	1.80
Waltham, Mass	19.20		16	M.	City	126	1.62
Washington, D. C.	31.00	1	25	A. N.	City	6,284	1.25

* Cost per M. M. Moonlight. Ex. Expired.
† Per Night. Mid. Midnight. N. C. No Contract.
A. N. All Night. Co. Company. Ind. Indefinite.

Examination and Valuation of Public Works.

When bonds based on the value or earning capacity of water, gas, electric or other public works are to be issued or such works are to be sold to municipalities or investors, it is desirable that they be subjected to expert examination, and that an estimate of their cost, present value and earning capacity be obtained as a basis for such bond issue or such sale.

In the past many over-issues of bonds have been made with the result that the purchaser has found his investment of little or no value.

Too great care cannot be taken to thoroughly investigate these securities before investing in them.

The writer will undertake to make such investigations and report on the value of such works.

Energy Lost in Electrical Transmission.

The energy lost in the transmission of electric currents varies with the size and arrangement of the wires. The loss in the transmission by direct currents is given by the formula:

$$L = \frac{10.5 \times l \times C}{A}$$

In this formula

L = loss in volts;
l = length of wire in feet;
C = the current in amperes;
A = the area of wire in circular mils.

The following diagram gives a graphical solution of such losses without computation.

WIRING DIAGRAM.

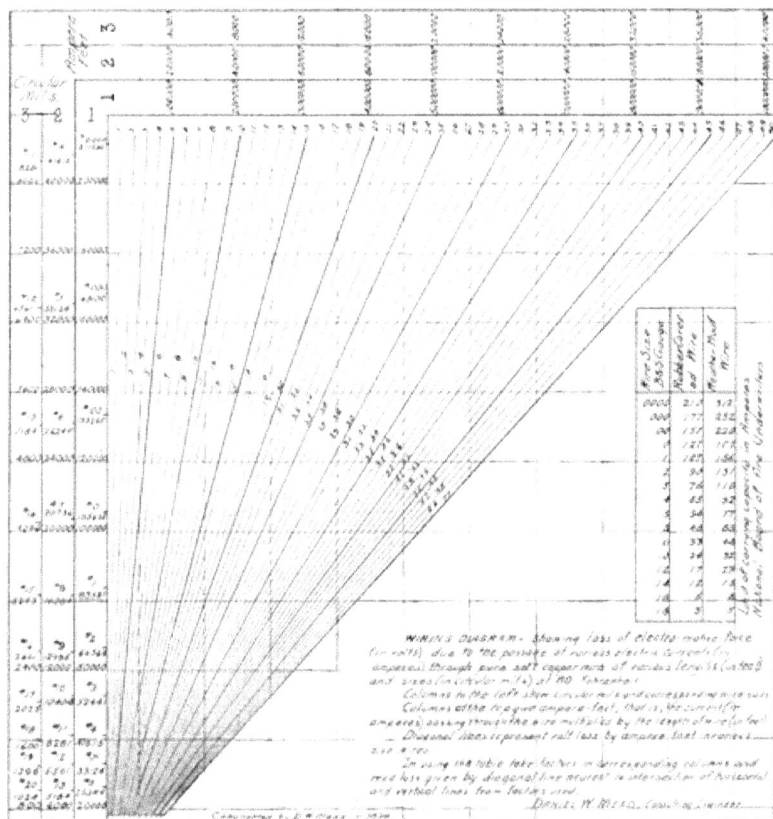

The Switchboard.

The Switchboard contains the instruments for operating the electrical plant. Much depends on its proper design and installation.

It is desirable to operate dynamos or motors singly or together, to cut one or another in or out of service without disturbing the operation of the balance of the plant; also to observe the fluctuation in energy use or voltage maintained. Careful design of the switchboard will add greatly to the facility and safety of operation of electrical plants.

The diagram on the opposite page illustrates a design of the writer's for the switchboard connections for two compound wound direct-current dynamos, to be operated singly or in multiple, on a 220-volt circuit, and operating either or both incandescent and arc lamps. The board as designed is in use in Lawrenceville, Ill., and is illustrated among the views of the Lawrenceville plant.

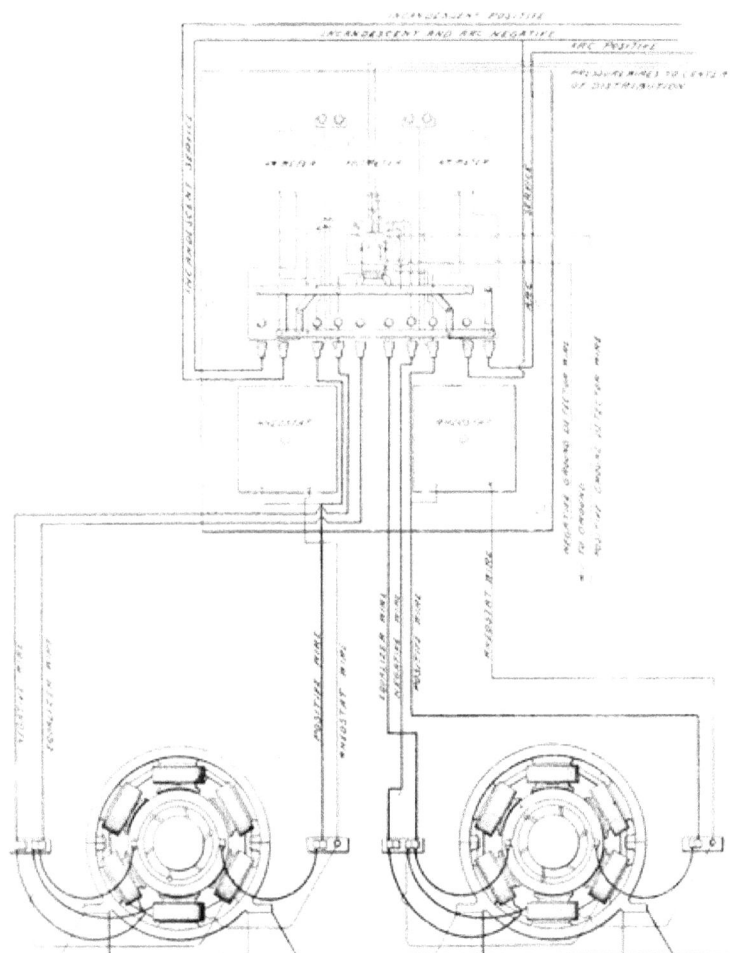

Switch Board Connections for Two Compound Wound Dynamos
in Multiple.

The Lawrenceville, Illinois, Water and Light Plant.

The following cuts illustrate a small electric light and water works installation designed by the writer for the Lawrenceville Light and Water Company. The engines and direct current generators are direct connected. The boilers are of an internally fired self-contained type. The water supply is obtained from deep wells, a portion of it being raised by special steam actuated deep well pumps, located at the central station, and a portion by an electric actuated deep well pump, located about one-half mile from the station, but controlled from the central station. The electric plant is of the 220 volt direct current constant potential system, and operates arc and incandescent lights and motors.

Power House—Lawrenceville Light and Water Co.

STEAM DRIVEN DEEP WELL PUMP.

ELECTRIC DRIVEN DEEP WELL PUMP.

ENGINE AND DYNAMO ROOM.

LAWRENCEVILLE LIGHT AND WATER CO.

Boiler Plant.
Switch Board.

Direct Connected Engine and Dynamo.
Lawrenceville Light and Water Co.

The DeKalb, Illinois, Electrical Pumping Plant.

The DeKalb Electrical Pumping Plant was designed and installed under the writer's supervision in 1894. It consists of a deep-well pump, to raise the waters of the St. Peter sandstone a distance of about 150 feet into a surface reservoir, and of two triplex power pumps to force the water from the reservoir into the water mains and stand-pipe. All pumps are actuated by electric motors, which are operated from the station of the De Kalb Electric Company, about one-half mile distant. For the year previous to the installation of this plant, the cost of pumping was 14 cents per 1,000 gallons. The cost of pumping since has been 4 cents per 1,000 gallons. This difference in cost of pumping forcibly illustrates the difference between good and bad engineering practice. It shows, also, what might be readily accomplished by the re-design and reconstruction of many of the existing plants now used, not only for waterworks but for other power purposes.

REAR VIEW.

FRONT VIEW.   ELECTRIC PUMPING PLANT   DEKALB, ILLINOIS.

The Examination and Improvement of Water Supplies.

The growth of population and the increase in manufacturing have both increased the demand for pure and adequate water supplies and at the same time, in many instances, created conditions which have rendered the supplies as they now exist both unsatisfactory and inadequate.

Comparatively few of the cities of the United States possess supplies that are entirely satisfactory, and numbers are constantly seeking new sources of supply or methods of increasing and improving present sources so that they will satisfy the demands of enlightened public opinion. Many thousands of deaths and much sickness is annually caused by water borne diseases which would be entirely prevented by proper sanitary precautions.

The question of the selection of a new supply or the improvement of present supplies is therefore a matter of the gravest importance. Water supply is a question wholly of local condition. No one source of supply nor no single method of development is universally applicable. Expert investigation is essential. The water resources should be closely examined and in many cases such examinations should be accompanied with chemical and bacteriological investigations. With such data at hand the expert can select the most desirable source of supply and the best method of improvement or development and results of the highest sanitary value may thus be attained.

The New Water Supply System at Rockford, Illinois.

During 1897 and 1898 the writer designed and constructed for the City of Rockford, Illinois, a water supply system of a capacity of 7,000,000 gallons per day.

The water is obtained by deep wells from the St. Peter and Potsdam sandstone and is organically pure, being equal if not superior to the supply of any city of its size (40,000). Much time and money had been expended in an attempt to furnish Rockford with an adequate supply of water but without success until the problem was satisfactorily solved by this construction.

The plant is essentially different in design from any plant hitherto installed. It consists of a vertical shaft 80 feet in depth, in which high grade centrifugal pumps are installed at the base, operated by vertical compound condensing engines at the surface. The power is transmitted from engine to pump by vertical manila rope drives. There are three engines of 150 H. P. each and three pumps, each capable of raising 6,000,000 gallons per day. From the base of the large shaft in which the pumps are set, is built a smaller shaft, from the bottom of which tunnels extend to the various artesian wells. Each well is piped to the common suction pipe of the pumps.

The entire work was constructed from 80 to 90 feet below the ground water level and for this reason pneumatic pressure was used. From 35 to 42 pounds pressure above atmosphere was carried for over eight months. The work was successfully completed under the writer's personal supervision under extremely discouraging conditions and after the work had been abandoned by the contractors as impracticable. Hon. E. W. Brown, Mayor, in describing this system before the Mayors of the Cities of Illinois, at Peoria, in the spring of 1899, said :

"This system which we have so successfully inaugurated is a radical departure from old plans and old methods. Its success has been so decided, however, that it is worthy of the most careful attention from any city within the artesian well belt that is desirous of increasing its water supply."

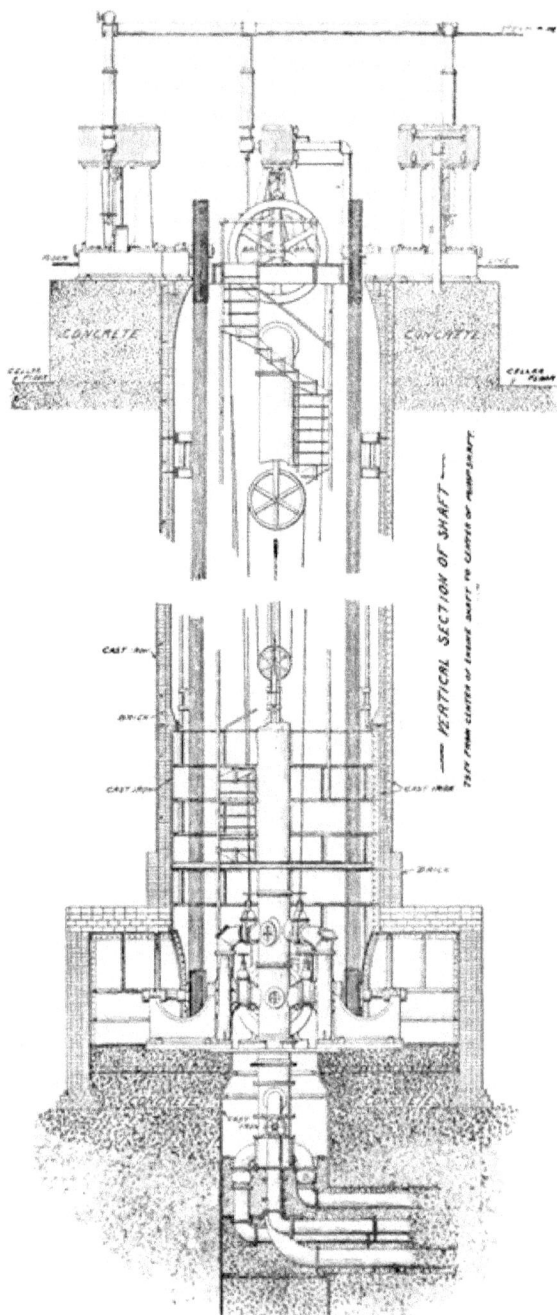

SECTION OF SHAFT AND ELEVATION OF PUMPS AND ENGINES.
PUMPING PLANT—ROCKFORD, ILLINOIS.

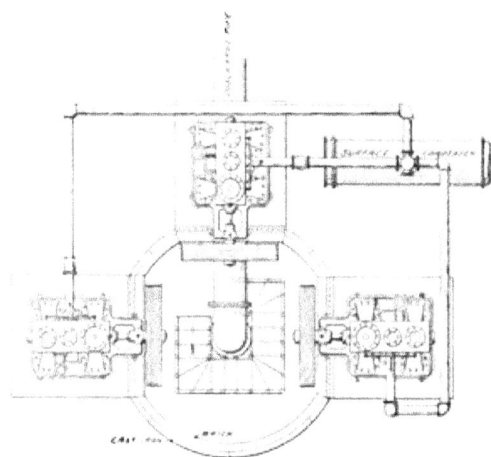

ARRANGEMENT OF ENGINES. PUMPING PLANT—ROCKFORD, ILL.

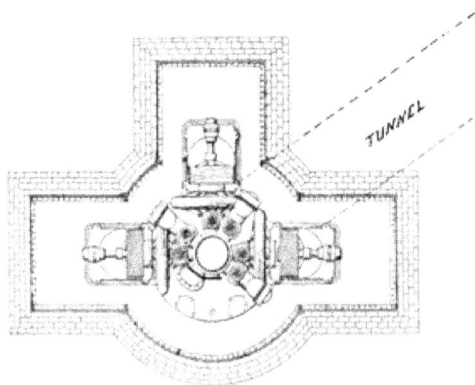

ARRANGEMENT OF PUMPS. PUMPING PLANT—ROCKFORD, ILL.

PUMP HOUSE—ROCKFORD, ILLINOIS.

SINKING THE CAISSON—THE LAST TEN FEET—ROCKFORD, ILLINOIS.

Sinking the Caisson—Rockford, Illinois.

COMPOUND CONDENSING ENGINES—ROCKFORD, ILLINOIS, PUMPING PLANT.*

CENTRIFUGAL PUMPS—ROCKFORD, ILLINOIS, PUMPING PLANT.*

*Cuts from *Engineering Record.*

Weiring the Water—Rockford, Illinois.
7,000,000 Gallons per Day.

STORAGE RESERVOIR—WEST DUNDEE, ILL.

Water Storage.

The cost of continuous pumping at rates of speed just sufficient to meet the demand, is usually much greater than where the pumping can be done at the most economical rate of the machinery and within a short period of time. This is especially true in the smaller places, where a few hours' pumping is sufficient to furnish water for the twenty-four hours, and where little water is used during the night. In such places some form of water storage is usually desirable.

In level localities, where the water in the lower portion of a stand-pipe could not be made available, stand-towers, consisting of metal tanks on steel or masonry towers, may often be used to advantage. Where elevations are available for storage sites, reservoirs built in or partially in the ground may be made available. With less elevation, stand-pipes can be used with good results. Where the desired amount of storage is small, pneumatic storage tanks can be installed with satisfactory results and often with much less expense. Each design may have its range of usefulness; none are universally desirable or applicable.

STAND TOWER—DUNDEE, ILLINOIS.

STAND PIPE ROCK ISLAND, ILLINOIS.

STAND TOWER—JERSEYVILLE, ILLINOIS.

PNEUMATIC STORAGE TANKS—SHABBONA, ILLINOIS.

Drainage.

Large tracts of land in the Upper Mississippi valley have already been reclaimed from overflow by means of ditches, levees and pumping.

The rapid increase in values due to such improvements has made this an attractive field for investment, which has proved not only profitable but also safe, when properly carried out.

The following cuts illustrate the improvement of the Meredocia Levee and Drainage District, near Albany, Illinois, carried out under the writer's direction.

The district is protected from inundation from both the Mississippi and Rock rivers by levees at the respective east and west ends of the district, and the storm-water falling on the district itself is removed by means of a direct-connected centrifugal pump of a capacity of 25,000 gallons per minute.

The success of these enterprises depends entirely on how well the improvements are designed and carried out with respect to the conditions, which differ largely with each case.

Power Station Meredocia Drainage District.

Engines and Pump.

Boiler Plant.

Meredocia Drainage District—Near Albany, Illinois.

The Levee—Meredocia Drainage District.

Weiring the Water—Meredocia Drainage District.
25,000 Gallons per Minute.

47

The Hydraulic Power Plant of the Rockford General Electric Co.

Electric power transmission has made possible the development of many water powers not hitherto available. Water powers as distant as eighty miles from manufacturing centers are now being successfully utilized for power purposes.

The successful installation of these combined hydraulic and electric power plants on an economical basis requires the most careful preparation of plans and attention to detail of construction, and is worthy of the best engineering service.

The following cuts illustrate the hydraulic-power plant of the Rockford General Electric Co. This plant was designed and constructed under the supervision of the writer. The wheel-pit was designed for ten 150 H. P. turbines, only four of which have yet been installed. The plant is used both for electric light and electric power transmission purposes.

Gate Post Bracing — Rockford General Electric Co.

Head Gates   Rockford General Electric Co.

WHEEL PIT—ROCKFORD GENERAL ELECTRIC CO.

WHEEL PIT—ROCKFORD GENERAL ELECTRIC CO.

Two Wheels in Place—Rockford General Electric Co.

Under the Wheel Pit—Rockford General Electric Co.

Hydraulic Machinery.

Hydraulic machinery is, under favorable conditions, the most satisfactory machinery for the generation and utilization of power, as it requires far less supervision and repairs than is necessary with other forms of power generators. The ordinary forms of turbine and impulse wheels are now furnished from stock by various companies who make a specialty of such goods.

Nevertheless, there is frequently a demand for special hydraulic appliances, which require special design and construction. The writer is prepared to furnish such designs and will be glad to correspond with parties having special or difficult conditions to be considered. A number of illustrations of hydraulic power and pumping plants have already been shown on previous pages. The cuts on the opposite page illustrate two forms of special hydraulic rams, designed by the writer. The ram is the simplest and most economical method of pumping water where hydraulic power is available. At Dundee, Illinois, is a hydraulic ram designed by the writer, which is about twenty feet in height and which furnishes water for that place. This ram takes water under a fifty foot head and delivers it about 100 feet above the pump house. The drive pipe is ten inches in diameter and over two thousand feet in length. It is believed to be the largest ram yet constructed and is similar in general design to the artesian well ram shown on the following page.

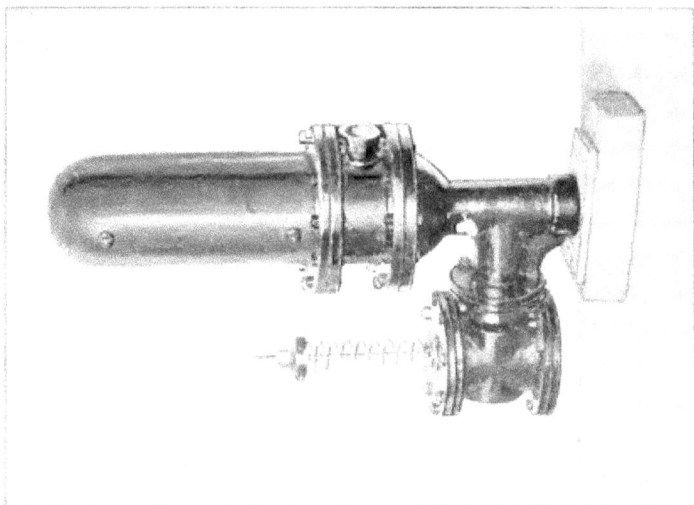

HYDRAULIC RAM FOR ARTESIAN WELLS.
DEER PARK, ILLINOIS.

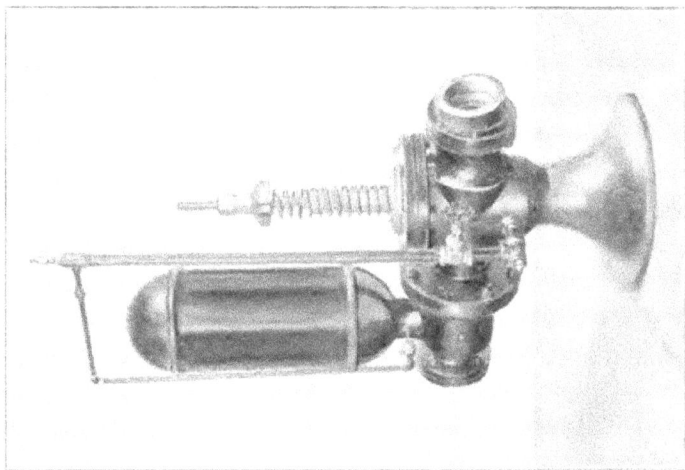

HYDRAULIC RAM FOR VILLAGE SUPPLY.
ALGONQUIN, ILLINOIS.

Foundation Work.

In this day of high buildings and large bridges, the matter of foundations must receive the most careful consideration. The writer has had an extended practical experience on foundation work, and will undertake the design or superintend the construction of any such work as requires special attention.

The cuts on the opposite page illustrate the construction of bridge foundations in the Rock river, at Rockford, Illinois, by the writer in 1890. The first cut illustrates piers constructed by means of coffer-dams. The second cut shows the open caisson method of construction.

The pneumatic method of construction is illustrated by the cuts on pages 36 and 37.

MORGAN STREET BRIDGE FOUNDATION— ROCKFORD, ILLINOIS.

STATE STREET BRIDGE FOUNDATION— ROCKFORD, ILLINOIS.

Sewers and Sewerage.

Modern conditions have made sewerage a necessary requirement of municipal growth. The correct design and construction of sewerage systems and sewage disposal works, is of the greatest importance for the Public Health.

Sewerage systems, like all other public works, must be modified to meet local conditions, and their success depends on the experience and judgment of their designer.

The following views illustrate a sewer machine at work on the Aurora, Illinois, sewers in 1892.

SEWER MACHINE AT WORK—AURORA, ILLINOIS.

SEWER MACHINE AT WORK — AURORA, ILLINOIS.

Special Machinery for Construction.

These cuts illustrate some special machinery for sewer work designed by the writer for the sewer work at Aurora, Illinois.

The advantage of labor-saving machinery needs but brief comment. To complete work at a profit, modern appliances are essential and such appliances must be simple and practicable. The writer's long experience in the practical construction of public works, has rendered him familiar with the possibilities of such machines and his services are available for the selection and installation of construction plants.

Notes on Pumps and Pumping.

All pumps should be set on solid foundations.

All steam or water pipes connecting with the pump should be as straight and free from bends as possible.

The suction pipe should be made thoroughly tight. A slight leak will seriously affect the suction lift.

Suction Lift—The limits of possible suction lift vary with the elevation of the locality above sea level. The available suction head at any place may be determined by the formula:

$$\text{Log } F = 1.53084 - \frac{H}{64000}$$

in which

F—height in feet for a column of water which will balance average atmospheric pressure.

H = height of the station above sea level.

About two feet must be deducted from this head for minimum atmospheric pressure, and a further deduction must be made for the loss in head in the suction pipe and suction chambers of the pump.

These various deductions leave the maximum possible suction lift from about twenty-seven feet at sea level to twenty-four feet, at an elevation of 1,500 feet above sea level.

It is usually desirable, however, to set pumping machinery nearer the water when possible.

Speed of Pumps—*Ordinary speed to run small pumps is 100 feet of piston speed per minute.* Many well designed pumps run at 300 feet or more.

Capacity of Pumps—*To find quantity of water elevated in one minute by a double acting pump running at 100 feet of piston speed per minute.* Square the diameter of water cylinder in inches and multiply by 4. Example:—Capacity of a five-inch cylinder is desired. The square of the diameter (five inches) is 25, and multiplied by 4 gives 100, which is gallons per minute (approximately).

Size of Pistons—*The area of the steam piston* multiplied by the steam pressure, gives the total amount of pressure that can be exerted. *The area of the water piston* multiplied by the pressure of water per square inch, the resistance. *A margin* must be made between the *power* and the *resistance* to *move* the pistons at the required speed—say from 20 to 50 per cent., according to speed and other conditions.

TABLE SHOWING THE CAPACITY OF PUMPING ENGINES.

Gallons in 24 hours	Gallons per hour	Gallons per minute	Piston Speed in feet per minute of each plunger of Duplex Pumps														
			8 inches	8½ x 9 inches	9 inches	10½ inches	11 inches	12 inches	13 inches	14 inches	15 inches	16 inches	17 inches	18 inches	19 inches	20 inches
250,000	10,416	173.6	43.4	33.2	29.4											
500,000	20,833	347.2	86.8	66.4	57.8	40.5	35.2	29.5								
750,000	13,250	520.8		99.3	88.2	60.7	52.8	44.3	37.8	32.6						
1,000,000	41,666	694.4			117.7	81.0	70.3	95.1	50.3	43.4	37.8	33.2				
1,250,000	52,083	868.0				100.1	87.9	73.8	62.9	84.3	47.2	41.6				
1,500,000	62,500	1,041.6					105.5	88.6	75.5	65.1	56.7	49.8	44.2			
1,750,000	72,919	1,215.2						103.4	88.1	76.0	66.2	58.2	51.5	45.9		
2,000,000	83,333	1,388.8							100.7	86.8	75.6	66.5	58.8	52.5	47.1	
2,250,000	93,749	1,562.4								113.2	97.7	85.0	74.8	66.2	59.1	53.47.2
2,500,000	104,166	1,736.0									108.6	94.4	83.2	73.5	65.6	58.9 53.2
3,000,000	125,000	2,083.3										113.4	99.7	88.3	78.8 70.7	63.8
3,500,000	145,833	2,430.5											116.4	103.0	91.9 82.5	74.4

Equivalent Measures and Weights of Water

At 4 Degrees Cent. or 39.2 Degrees Fahr.

U.S. Gallons	Cubic Feet.	Cubic Inches.	Imperial Gallons.	Liters.	Cubic Meters.	Pounds.
1	.13368	231.	.83321	3.7853	.0037853	8.34112
7.48055	1	1728	6.23287	28.3161	.0283161	62.3961
.004329	.0005787	1	.003607	.0163866	.0000164	.0361089
1.20017	.160439	277.274	1	4.54303	.00454303	10.0168
.264179	.035316	61.0254	.22042	1	.001	2.20355
264.179	35.31563	61025.4	220.117	1000	1	2203.55
.119888	.0160266	27.694	.099892	.453813	.000453813	1

Constants for Calculating the Friction of Water in Iron Pipe.

| Diam. of Pipe. | | U.S. gallons per foot of length. | C=60A. | | K=Constant for pipe 100 ft. long. | |
Inches.	Feet. d.	A.	Number. C.	Logarithm.	Number. K.	Logarithm.
3	.250	.37	22.0	1.34305	.16514	9.21784
4	.333	.65	39.2	1.59293	.11608	9.06476
5	.417	1.02	61.2	1.78675	.08914	8.95005
6	.500	1.47	88.1	1.94511	.07221	8.85858
7	.583	2.00	119.9	2.07901	.06062	8.78264
8	.667	2.61	156.7	2.19499	.05221	8.71778
9	.750	3.30	198.3	2.28730	.04376	8.64111
10	.833	4.08	244.8	2.38881	.04084	8.61107
11	.917	4.94	296.2	2.47160	.03682	8.56606
12	1.000	5.87	352.5	2.54717	.03351	8.52523
13	1.084	6.89	413.7	2.61670	.03075	8.48788
14	1.167	8.00	479.8	2.68107	.02841	8.45346
15	1.250	9.18	550.8	2.74099	.02640	8.42156
16	1.333	10.44	626.7	2.79705	.02465	8.39182
18	1.500	13.22	793.1	2.89936	.02177	8.33781
20	1.667	16.32	979.2	2.99087	.01949	8.28974
24	2.000	23.50	1410.0	3.14923	.01611	8.20709
30	2.500	36.72	2203.2	3.34305	.01278	8.10668
36	3.000	52.88	3172.6	3.50142	.01060	8.02514
42	3.500	71.97	4318.2	3.63531	.00905	7.95650
48	4.000	94.00	5640.1	3.75129	.00789	7.89731

$H = lKv^2$

$h = \dfrac{v^2}{2g} = .0155v^2$

$Q = Cv$ (see table).

$v = \dfrac{Q}{C}$

$C = 60A$ (see table).

A=U.S. gallons per ft. of pipe.

Q=discharge in U.S. gals. per min.

g=acceleration due to gravity.=32.16.

l=length of pipe in feet.

v=velocity per second in feet.

d=diameter of pipe in feet.

h=velocity head in feet.

H=friction head in feet.

K=constant from D'Arcy's formula.

Weir Formulas.

For calculating the flow of water over weirs the following formulas are in use:

For sharp crested weirs without end contractions:

1. Francis, $Q = 3.33 \ L \ H^{\frac{3}{2}}$

2. Fteley & Stearns, $Q = 3.31 \ L \ H^{\frac{3}{2}} + .007 \ L$

For broad crested weirs without end contractions:

3. Francis, $Q = 3.01 \ L \ H^{1.53}$

4. Union, $Q = 3.09 \ L \ H^{\frac{3}{2}}$

To compensate for a single end contraction in a long weir, deduct from the total length in feet an amount equal to one-tenth the head upon the weir in feet. Reduce the total length a like amount for each end contraction. With end contractions the Francis formula becomes:

5. $Q = 3.33 \ (L - \frac{H}{10} \ N) \ H^{\frac{3}{2}}$

If there is velocity of approach, divide the weir volume as above by section of channel in square feet for approximate velocity, V. Then the additional depth on weir due to this velocity is $h = [V^2 \div 64.4]$. Add to the measured depth 1.5 h for the corrected depth on weir.

H = Head of water on weir in feet; L = Length of weir in feet;
N = No. of end contractions; Q = disch. in cu. ft. per. sec.;
h = velocity head in feet; V = velocity of approach in feet.

Table of the $\frac{3}{2}$ Power of Numbers from .01 to 3.05.
For Use in Weir Formulas.

H.	.00	.01	.02	.03	.04	.05	.06	.07	.08	.09	D.
.0	.0000	.0010	.0028	.0052	.0080	.0112	.0147	.0185	.0226	.0270	46
.1	.0316	.0365	.0416	.0469	.0524	.0581	.0640	.0701	.0764	.0828	66
.2	.0894	.0962	.1032	.1103	.1176	.1250	.1326	.1403	.1482	.1562	81
.3	.1643	.1726	.1810	.1896	.1983	.2071	.2160	.2251	.2343	.2436	94
.4	.2530	.2625	.2721	.2819	.2919	.3019	.3120	.3222	.3325	.3430	106
.5	.3536	.3643	.3751	.3859	.3968	.4079	.4191	.4304	.4417	.4533	115
.6	.4648	.4764	.4882	.5001	.5120	.5240	.5362	.5484	.5607	.5732	125
.7	.5857	.5983	.6110	.6238	.6366	.6495	.6626	.6757	.6889	.7022	133
.8	.7155	.7290	.7426	.7562	.7699	.7837	.7975	.8114	.8254	.8396	142
.9	.8538	.8681	.8825	.8969	.9114	.9259	.9406	.9553	.9701	.9850	150
1.0	1.0000	1.0150	1.0301	1.0453	1.0606	1.0759	1.0913	1.1068	1.1224	1.1380	157
1.1	1.1537	1.1695	1.1853	1.2012	1.2172	1.2332	1.2494	1.2655	1.2818	1.2981	164
1.2	1.3145	1.3310	1.3475	1.3641	1.3808	1.3975	1.4143	1.4312	1.4482	1.4652	170
1.3	1.4822	1.4994	1.5166	1.5339	1.5512	1.5686	1.5860	1.6035	1.6211	1.6388	177
1.4	1.6565	1.6743	1.6921	1.7100	1.7280	1.7460	1.7641	1.7823	1.8005	1.8188	183
1.5	1.8371	1.8555	1.8740	1.8925	1.9111	1.9297	1.9484	1.9672	1.9860	2.0049	189
1.6	2.0239	2.0429	2.0620	2.0811	2.1002	2.1195	2.1388	2.1581	2.1775	2.1970	195
1.7	2.2165	2.2361	2.2558	2.2755	2.2952	2.3150	2.3349	2.3549	2.3749	2.3949	201
1.8	2.4150	2.4351	2.4553	2.4756	2.4959	2.5163	2.5367	2.5572	2.5777	2.5983	207
1.9	2.6190	2.6397	2.6605	2.6813	2.7021	2.7230	2.7440	2.7650	2.7861	2.8072	212
2.0	2.8284	2.8497	2.8710	2.8923	2.9137	2.9352	2.9567	2.9782	2.9998	3.0215	217
2.1	3.0432	3.0650	3.0868	3.1087	3.1306	3.1525	3.1745	3.1966	3.2187	3.2409	222
2.2	3.2631	3.2854	3.3077	3.3301	3.3525	3.3750	3.3975	3.4202	3.4428	3.4654	227
2.3	3.4881	3.5109	3.5337	3.5566	3.5795	3.6025	3.6255	3.6486	3.6717	3.6948	233
2.4	3.7181	3.7413	3.7646	3.7880	3.8114	3.8349	3.8584	3.8820	3.9056	3.9292	236
2.5	3.9528	3.9766	4.0014	4.0242	4.0481	4.0720	4.0960	4.1200	4.1441	4.1682	242
2.6	4.1924	4.2166	4.2409	4.2652	4.2895	4.3139	4.3383	4.3628	4.3873	4.4119	247
2.7	4.4366	4.4612	4.4859	4.5107	4.5355	4.5604	4.5853	4.6101	4.6351	4.6602	251
2.8	4.6853	4.7104	4.7356	4.7608	4.7861	4.8114	4.8367	4.8621	4.8875	4.9130	255
2.9	4.9385	4.9641	4.9897	5.0154	5.0411	5.0668	5.0926	5.1184	5.1443	5.1702	260
3.0	5.1962	5.2222	5.2483	5.2744	5.3005	5.3266					

The following table is calculated by the Francis formula for sharp crested weirs:

Discharge of Rectangular Weir — No End Contractions.

Given in Cubic Feet per Second for a Weir 1 Foot Wide for each .01 Foot in Depth from .20 to 2.09 Feet.

Depth in feet.	.00	.01	.02	.03	.04	.05	.06	.07	.08	.09
.2	0.2978	0.3205	0.3436	0.3673	0.3915	0.4162	0.4415	0.4672	0.4934	0.5200
.3	0.5472	0.5748	0.6028	0.6313	0.6602	0.6895	0.7193	0.7495	0.7800	0.8110
.4	0.8424	0.8742	0.9064	0.9390	0.9719	1.0052	1.0389	1.0730	1.1074	1.1422
.5	1.1773	1.2128	1.2487	1.2849	1.3214	1.3583	1.3955	1.4330	1.4709	1.5091
.6	1.5476	1.5865	1.6257	1.6652	1.7050	1.7451	1.7855	1.8262	1.8673	1.9086
.7	1.9503	1.9922	2.0344	2.0770	2.1198	2.1629	2.2063	2.2500	2.2940	2.3382
.8	2.3828	2.4276	2.4727	2.5180	2.5637	2.6096	2.6558	2.7022	2.7490	2.7959
.9	2.8432	2.8907	2.9385	2.9865	3.0348	3.0831	3.1322	3.1813	3.2306	3.2802
1.0	3.3300	3.3801	3.4304	3.4810	3.5318	3.5828	3.6342	3.6857	3.7375	3.7895
1.1	3.8418	3.8943	3.9470	4.0000	4.0532	4.1067	4.1604	4.2143	4.2684	4.3228
1.2	4.3774	4.4322	4.4873	4.5426	4.5981	4.6538	4.7098	4.7660	4.8224	4.8790
1.3	4.9358	4.9929	5.0502	5.1077	5.1654	5.2233	5.2814	5.3398	5.3984	5.4572
1.4	5.5162	5.5754	5.6348	5.6944	5.7542	5.8143	5.8745	5.9350	5.9957	6.0565
1.5	6.1176	6.1789	6.2404	6.3020	6.3639	6.4260	6.4883	6.5508	6.6135	6.6764
1.6	6.7394	6.8027	6.8662	6.9299	6.9937	7.0578	7.1221	7.1865	7.2512	7.3160
1.7	7.3810	7.4463	7.5117	7.5773	7.6431	7.7091	7.7752	7.8416	7.9081	7.9749
1.8	8.0418	8.1089	8.1762	8.2437	8.3113	8.3792	8.4472	8.5154	8.5838	8.6524
1.9	8.7212	8.7901	8.8592	8.9285	8.9980	9.0677	9.1375	9.2075	9.2777	9.3481
2.0	9.4187	9.4894	9.5603	9.6314	9.7026	9.7741	9.8457	9.9174	9.9894	10.0620

Catchment Basin Formulas.

For drainage the following formulas for maximum flood discharge have been proposed:

Fanning, $Q = 200 \, A^{\frac{5}{6}}$

Dredge, $Q = 1,300 \, \dfrac{A}{L^{\frac{2}{3}}}$

Cooley, $Q = 180 \, A^{\frac{3}{4}}$

The following formulas have been proposed for use in sewer design:

Hering, Gray & Stearn, $q = 5.89 \, A^{\frac{3}{4}}$

Talbot 1. $q = \dfrac{55}{a^{\frac{1}{2}} + 20}$

Talbot 2. $q = \dfrac{28}{a^{\frac{1}{2}} + 13}$

Q = max. flood discharge cu. ft. per sec.
q = max. flood discharge cu. ft. per sec. per acre.
A = area of watershed in sq. miles.
a = area of watershed in acres.
L = Extreme length of watershed.

Fire Streams.

The following table and data are compiled from the experiments of J. R. Freeman, C. E., as described in a paper read before the American Society of Civil Engineers, November, 1889, entitled, "Experiments Relating to Hydraulics of Fire Streams."

Table of Effective Fire Streams.

Using Smooth Nozzle and 100 Feet of 2 1-2 Inch Ordinary Best Quality Rubber-Lined Hose Between Nozzle and Hydrant, or Pump.

Nozzle Size	¾-inch.						⅞-inch.						1-inch.					
Pressure at Hydrant, lbs.	32	43	54	65	75	86	34	46	57	69	80	91	37	50	62	75	87	100
Pressure at Nozzle, lbs.	30	40	50	60	70	80	30	40	50	60	70	80	30	40	50	60	70	80
Pressure lost in 100 ft. 2½ in Hose, lbs.	2	3	4	5	5	6	4	6	7	9	10	11	7	10	12	15	17	20
Vertical Height, ft.	48	60	67	72	76	79	49	62	71	77	81	85	51	64	73	79	85	89
Horizontal Distance, ft.	37	44	50	54	58	62	42	49	55	61	66	70	47	55	61	67	72	76
Gals. Discharged per minute	90	104	116	127	137	147	123	142	159	174	188	201	161	186	208	228	246	263

Nozzle Size	1⅛-inch.						1¼-inch.						1⅜-inch.					
Pressure at Hydrant, lbs.	42	56	70	84	98	112	49	65	81	97	113	129	58	77	96	116	135	154
Pressure at Nozzle, lbs.	30	40	50	60	70	80	30	40	50	60	70	80	30	40	50	60	70	80
Pressure lost in 100 ft. 2½ in. Hose, lbs.	12	16	20	24	18	32	9	25	31	37	43	49	28	37	46	56	65	74
Vertical Height of Stream, ft.	52	65	75	83	89	92	53	67	77	85	91	95	55	69	79	87	92	97
Horizontal Dis. of Stream, ft.	50	59	66	72	77	81	54	63	70	76	81	85	56	66	73	79	84	88
Gals. Discharged per minute	206	238	266	294	314	336	256	296	331	363	392	419	315	363	406	445	480	514

The heights and distances given for good "effective fire streams" are with moderate wind.

Maximum vertical height reached by the spray or drops in still air, is from 22 per cent. for the lower pressures, to 56 per cent. for the higher pressures, higher than the elevations given in the table. Maximum horizontal distance reached by the spray or drops in still air, is about 120 per cent. for the lower pressures and 150 per cent. for the higher, further than the distance given in the table.

When "unlined linen hose" is used, the friction or pressure loss is from 8 to 50 per cent., increasing with the pressure. "Mill hose" is better than unlined linen hose for long lengths, but the ordinary best quality, smooth, rubber-lined hose is superior to the "mill hose," having less frictional resistance.

The "ring nozzle" is inferior to the smooth nozzle and actually delivers less water than the smooth. For instance, a ⅞-inch ring nozzle discharges the same quantity of water as a ¾-inch smooth, and a 1-inch ring nozzle the same as a ⅞-inch smooth.

Use double lines of hose and a Siamese nozzle for a long distance and a hot fire. A double line a thousand feet long delivers a 1¼-inch stream with the same force as a single line 287 feet long. Small streams are all right for small fires, but for large, hot fires use a 1¼-inch or a 1⅜-inch stream. Such a stream will always make a black mark wherever it hits, and the stream which hits and cools the burning coals is the "effective fire stream." Small streams are converted into steam before touching the coals.

FINALE.

The preceding pages can give only a general idea of a few phases in the design and construction of public works.

Each problem must have separate consideration and a solution of its own if the best results are desired.

Skilled investigation and careful study into local conditions and resources will lead to intelligent planning of improvements and in the end will give the best results—that is, the greatest return on the investment.

The value of carefully matured and thoughtfully made plans, and skillfully constructed work, are not always understood or appreciated in the beginning.

Experience, however, is a dear teacher and the value of such design and construction is fully understood in the end.

www.ingramcontent.com/pod-product-compliance
Lightning Source LLC
Chambersburg PA
CBHW022010190326
41519CB00010B/1464